CONTENTS

HOW TO MAKE A SILENCER FOR A MINI-14

PALADIN PRESS • BOULDER, COLORADO

How to Make a Silencer for a Mini-14

ISBN 0-87364-913-3
Printed in the United States of America

Published by Paladin Press, a division of
Paladin Enterprises, Inc., P.O. Box 1307,
Boulder, Colorado 80306, USA.
(303) 443-7250

Direct inquiries and/or orders to the above address.

WARNING

Since 1934, silencers have been legal in the United States only if they are registered to their owner and a $200 tax has been paid on their transfer. Current federal statutes allow a maximum penalty of 10 years in prison and a $10,000 fine for the illegal manufacture or possession of an untaxed, unregistered silencer for a portable firearm. The agency of enforcement of these federal laws is the Bureau of Alcohol, Tobacco, and Firearms of the U.S. Treasury Department. Inquiries regarding the procedures for applying for a license to construct a silencer and for paying the appropriate tax should be directed to that office.

In addition, many state and local jurisdictions have restrictions on the possession of silencers, even if appropriately taxed and registered with federal authorities.

Finally, technical data presented here, particularly on the manufacture, alteration, adjustment, and use of firearms and firearm silencers, inevitably reflect the author's individual beliefs and experience with particular firearms, equipment, and components under specific circumstances that the reader cannot duplicate exactly. The information in this book should therefore be used for guidance only and approached with great caution. Neither the author, publisher, or distributors of this book assume any responsibility for the use or misuse of information contained in this book.

This book is presented for academic study only.

PREFACE

I need to expand and emphasize this "fair warning" part of my book. The events at Waco, Texas, should have made everyone aware of what I already knew about the Feds. They are power mad! They *will* fire upon women and children, and so will the military! Friend, I *guarantee* that if your "group" has five people in it, one of them is, or will become, a snitch for the government. I am warning you most emphatically! Tyrants fear and hate silenced firearms (when those arms are not in the hands of their minions). Beware, beware, beware!

The manufacture or possession of firearm silencers, suppressors, sound modulators, or whatever you want to call them is a violation of federal and some states' laws, and it *will* get you serious prison time. If you already have a felony record, being caught with a silencer means that you'll serve 24 years in prison. *This book is for academic study only!*

Paladin Press has published another of my books entitled *How To Make A Silencer For A .22*. In it, I go into considerable detail about the basics of silencer construction. If you have not read it, you should do so. It's much better to make your mistakes on a $50 (used) single-shot .22 and at 10,000 psi (pounds per square inch) rather than on a more expensive Mini-14 generating 48,000 psi!

In this book, I'm going to assume that you can handle a MIG welder, a TIG welder, and a lathe. If you cannot, you are playing with nitroglycerin if you try to follow the instructions in this book. Moreover, you must read *all* of this book (and understand it all) before attempting any work!

HOW TO MAKE A SILENCER FOR A MINI-14

I do not consider the Mini-14 to be the ultimate survival rifle. The 11-inch-barreled CAR-15 has that honor. When a used CAR was $500 and a used Mini-14 was $300, there was no reason to settle for second best. However, since legislation has outlawed further manufacture of AR-15s, used Mini-14s go for $600, and CARs can go as high as $1,600! Few of us have that much desire to own a survival rifle! Also, there's the matter of $250 worth of scope mount and scope, $150 luminous inserts for the iron sights, $100 for a trigger job, $100 for a folding buttstock, $200 for a .22 Long Rifle conversion unit, plus enough magazines, hunting softpoints, and full-metal jacket military type ammo to make a difference in your skill level.

I must take this opportunity to *again* warn you that the manufacture or possession of silencers is a federal felony unless you are in possession of a Title II or III Federal Firearms License (FFL). If you are not so licensed, then this book is for academic study only! Beware! If your "friends" are caught doing *anything* illegal, they will tell the cops everything they know, including the fact that you make "cans" for guns.

For those who are properly licensed to make this silencer, the most troublesome parts are getting the barrel threaded and arranging for a new front sight. Take some notes on how the AR-15 front sight bracket is mounted on its barrel. You cannot use an AR-15 front sight on the Mini-14, however, because it stands much too high above the barrel. You'll have to make and mount a similar one in order to have your Ruger's front sight be the same height above the bore line as it is now (while allowing for the increase of barrel diameter from 1/2 inch or so at the muzzle to 5/8 inch at the forend). A gunsmith will charge you nearly $200 for such a job, so I suggest you contrive a way to do it yourself.

Be aware that silver-soldering or welding a sight onto a barrel can ruin the barrel! If you use welder's heat control paste in the bore and have the sight base already tinned with a coating of silver-solder, it is possible to avoid getting scale in the bore. You can cast a lead "lap" onto a bronze bristle bore brush. Put lapping paste on the lap (which has rifling grooves in it, because you cast it *in* the muzzle), use a cleaning rod to move this lap through the bore, and polish out any roughness that you may have created by heating the barrel. You can also (carefully!) drill and tap a pair of shallow .060 (60/1000) inch holes in the barrel and mount the ramped-style sight with some 6 x 48 scope mount set screws. However, the AR-15 style of bracket mount is the best way to install a sight on your Mini-14.

Removing and replacing a barrel is a big job on most high-powered rifles. Gunsmiths get $200 or so to do it, and for good reason. The headspace on a .223 rifle must be perfect or you risk getting hot gases and brass fragments in your face. The .223 operates at about 48,000 psi (a garage's air nozzle has about 150 psi), so you've got a lot of pressure right under your nose. Also, semiautomatic rifles are intended for rapid fire use in combat, so nothing less than perfect functional reliability is acceptable! The barrel must be threaded in such a way that when the rear of the barrel's shoulder comes up against the front of the receiver, the front sight is properly aligned and the headspace is correct. Headspace is the space that exists between the cartridge, the front of the chamber, and the bolt face. Too much headspace can let the brass case rupture under the firing pressures. Not enough headspace means that the bolt can't close properly on a chambered round of ammo. Either of these situations can result in your face getting altered, so neither one can be allowed to exist! A gunsmith employs two gauges (a "go" and a "no-go") to check for inadequate or excessive headspace.

You also need $200 worth of specialized tools for barrel removal (a barrel vise and an action wrench). The CAR-15 comes with an already threaded muzzle on its 11 1/2 inch barrel. Federal regulations hold that such a short barrel must have a 5 1/2 inch flash hider and that it must be welded in place (in addition to the threads by which it is mounted on the barrel). This weld sometimes hinders accuracy, as can an improperly aligned flash hider (or silencer). A Dremel (hand-held) grinder and its carborundum cut-off disks can slice through this weld, allowing you to screw the flash hider off, and permitting attaché case or backpack concealment of the disassembled CAR-15 (or Mini-14). *Know that by doing this, you would be risking a 10-year sentence in federal prison!*

There are ways to have a removable flash hider that appears not to be. Loctite thread sealer can prevent casual unscrewing, yet the heat of a butane lighter will break its bond on the threads. Gun-Kote (a type of spray-on coating) replicates the look of the military Parkerizing that is a commonly encountered finish on guns. Naturally, several types of expoxies will fill in the slot that the Dremel cut-off disk will have made in the welds that secure the flash hider from being unscrewed, and after the "Gun-Kote" is applied, the epoxied-in slot can no longer be seen (yet the epoxy will melt from a butane lighter's flame, so the flash hider can still be screwed off). This is a risky endeavor though, since you may be certain that federal agents will have read this book! Once again, I do not advocate that anyone break *any* laws! I am just attempting to foresee problems and describe options as you evaluate your choice of survival rifles.

The reason you can't cut a Mini-14's barrel shorter than 14 inches is because the Ruger has a gas piston and gas cylinder concealed in the forestock. When you cut the barrel too short, you change the gas pressure that is used to function the action, so you'll either have to enlarge the gas port into the barrel (a tricky job) or leave the barrel longer.

In this book, I am assuming that you wish to retain the original 18 1/2 inch length of your Ruger's barrel. I will show you a way in which you can achieve a threaded end on your barrel without removing the barrel from the receiver or taking the barrel to a machine shop. This method does require precision machining, though! You must be able to use a micrometer and measure the outside diameter (O.D.) of your rifle's muzzle to within .0005 (1/2 of 1/1000 of an inch). The "press fit" or "interference fit" method shown herein is used to mount wheel bearings on car's rear axles, so it is a solid arrangement, but it does require close attention to detail. If you make the hole in the threaded adapter (Fig. 1) too small, and I mean even .002 inches too small, you can blow up your rifle! This is because the too-tight fit of the adapter around the barrel will be constricting the bore through which the bullet passes. This constriction will cause the working pressures of the powder gases to increase, and with a .223, they are already at the maximum level commensurate with safe use! You'll need to make a simple sketch of what you want and find a machine shop that has both a precision lathe and an internal honing machine. In making the adapter, the machinist will use the lathe to get close to the final measurements, but then he will use the hone to remove the final 2-3 thousandths of an inch of metal necessary to achieve your specified size. Most barrels taper a bit, so you must measure at both the muzzle and 1 1/4 inch back from the muzzle.

If you make the hole in the adapter too large for the barrel, it will be too loose, and it will not hold the silencer in proper alignment with the bore. The fired bullets will then strike the baffles inside the can, and that's bad news! All my can designs incorporate a sort of "safety valve" front end cap, which releases when excessive pressures exist in the can. So although the silencer cannot blow up at the end of your barrel, the misalignment will cause you to miss your target, and any subsequent shots will be loud because the pressures will have blown the guts out of the can. Without the baffles to absorb the noise-producing heat from the gases, the silencer won't work!

Let's say you "mike" the outside of your barrel and get 1/2 inch (.5000 inch) exactly. You then tell the machine shop that you want a hole bored lengthwise (1 1/8 inch long) through a 7/8 inch by 14 bolt, and you want that hole to have an inside diameter (I.D.) of .4985-.4995 inch. You are then allowing him a tolerance of a 1/2 of 1/1000 of an inch on either side of .499 inch. That .499 inch is .001 (1/1000) inch smaller than the O.D. of your .500 (1/2) inch barrel (Fig. 1). You need the adapter's hole to be .003 (3/1000) inch larger than the barrel's O.D. in order for you to be able to slip the adapter onto the barrel. You achieve this clearance by clamping the barrel (in a padded vise) and "shrinking" its O.D. by .001 (1/1000) inch or so and by making the adapter expand about .004 (4/1000) inch.

To do this, simply spray freon or butane on the end of the barrel and "play" an oxyacetylene welding torch on the adapter. The freon's cold will shrink the barrel, and the torch's heat will make the adapter expand. Hold the rear flange of the adapter in a pair of pliers, and have a hardwood mallet at hand. When the adapter is a uniform dull-red in color, quickly tap it into place on the barrel. When the temperatures have returned to normal, the threaded adapter will be "frozen" onto your barrel, and you now have an external (male) thread on which the female thread of the can will be screwed.

You will not be able to remove this adapter short of using the Dremel and cut-off disks to slit it in two places (180 degrees apart) and pulling them off of your barrel. So be aware that you'll also need to make and mount a homemade flash hider unless you're willing to have those big, coarse, 14 TPI (threads per inch) seen by the public. If so, the Bureau of Alcohol, Tobacco, and Firearms (BATF) *will* search your house, car, and business in an attempt to find the can that they'll be sure you have! By the way, since your 18-inch barrel is over the minimum legal length, the flash hider need not be welded into place in order to be legal. However, I say again, the 1994 Crime Bill makes it illegal to manufacture an "assault rifle," and if your Mini-14 has a folding buttstock, threading the barrel's front is a violation of this law. Don't say you weren't warned, because I am warning you now!

Ideally, you will first test-fire the now-threaded barrel by firing a round through a .22 LR conversion unit. I would fire even the .22 by remote control at first, because I value my eyes and fingers! Tie the gun to a tree, car axle, or other stable support and pull the trigger with a string. Do this several times. Assuming all is well, fire some more rounds while holding the rifle horizontally down near your hip. Naturally, never fire any gun without protective glasses and earmuffs in place.

Here is a good reason to already have your new front sight in place *before* adding the threaded adapter. If the .22 unit doesn't group as tightly on the target as it did before installing the threaded adapter, you've done something wrong. Of course, if you have a scope and mount for your Ruger, you can check accuracy without mounting the new front sight.

You should definitely follow these safety firing tests before unleashing the .223's nearly 50,000 psi pressure just inches from your eyes! *Do not* proceed with making the can before you've tested the barrel adapter's suitability. If you expect any accuracy at all, a silencer for a high-powered rifle must be aligned perfectly with the bore, and it must be mounted very rigidly. A powerful rifle's can must be large and heavy, and that complicates things (as compared to a can for a pistol or a .22 rifle).

This is why I arrange a two-point mount for this Mini-14 can. A large silencer mounted on the end of a long, thin rifle barrel can "sag" enough to cause the bullet to strike the silencer's baffles. The larger the hole is through the center of a can, the more powder gases can eventually emerge, resulting in noise. However, the closer the diameter of the hole is to the diameter of the bullet, the less play or tolerance you have for any misalignment of the can with the barrel's bore. The hole size through the baffle in this can is 1/4 inch (.250 inch). The diameter of the bullet is .224 (224/1000) inch. That means you have only .013 (13/1000) inch of clearance around the bullet as it passes through the baffles. A fine hair is close to that size, so do you see why there's a need for additional rigidity in the mount?

The threaded end plug is set 7 inches into the can (from the rear) and welded into the tube (Fig. 2). The rear 3/8 inch of the 1 1/8 inch O.D. tube is slitted lengthwise and compressed around a 5/8 inch thick neoprene disk by a hose clamp (Figs. 3 and 4). Some of the powder gases can expand back into this area (where the rear 7 inches of the can are "telescoped" back around the front 7 inches of your Ruger). The compressed neoprene disk achieves a gas-tight seal at the can's rear, so you can utilize this area (between barrel and can) to vent some of the powder gases, thus making the silencer more efficient.

The additional support of this two-point mounting will be appreciated if the can ever receives any sort of a blow (like if you stand the rifle on its butt against a tree and it gets knocked over). If something like this should ever happen, you'd damn well better carefully test-fire the piece again before putting it back into "duty" status. A silencers' fragility is one of the main reasons why the military doesn't have much use for them. The vast majority of soldiers are just far too careless with their gear and so cannot be trusted with something that is both so vital and so easily screwed up!

Cans for high-powered rifles must be "sleeved" (Fig. 5), and they must have an expansion chamber arranged inside their rear (Fig. 6). If you try to just pack the can with my compressed screen wire "doughnut" type of baffle (as depicted in *How to Make a Silencer for a .22*), the greater heat and blast of the more powerful .223 rounds will destroy the baffles in just a few shots. Making and mounting this can is a lot of work, so do not settle for not being able to fire hundreds of rounds in between the necessary cleanings of the silencer.

The expansion chamber is nothing more than a 4 inch long steel spool around which you loosely wrap stainless steel screen wire. The smaller of the two tubes that make up the can (1 1/8 inch diameter) has holes drilled through its sides at its rear (see Fig. 7). Thus, the hottest, highest-pressure gases are vented immediately into the area between the two tubes (the sleeving effect), and this diversion of gases helps to lighten the job of heat dissipation that the baffles in the front 6 inches of the can must accomplish.

The spool is just a 4-inch long piece of 1/2 inch O.D. steel tubing with .095 (95/1000) inch thick walls. This means that the hole through this tube measures .310 (310/1000) inch. That is .043 (43/1000) inch clearance all around the .224 (224/1000) inch bullet, but this greater tolerance is necessary! The ends of this spool are made of two washers, one of which is welded around each end of the 4 inch long tube. These washers must be precision-turned on a lathe so as to maintain concentricity. Their I.D. must be cut with a boring bar to .502-.504 inch so that they slip over the 4-inch long tube. Their O.D. must be turned first, however. They must slip through the front 10 inches of the inside tube (the inside diameter of which is .995 inch). The O.D. of the washers needs to be .990 (990/1000) inch so that the finished spool can be installed inside the can.

You can reduce a washer's O.D. by using a nut and lockwasher to lock the washer onto a bolt. Chuck the threaded end of this bolt into the lathe and center-drill the bolt head. Move the bolt further out of the chuck (to give the lathe's cutter access to the washer's O.D.) and bring the tailstock's live center into the cut you've center-drilled into the bolt head. Proceed to use high RPMs and light cuts to reduce the washer's O.D. to the desired size. These measurements cannot be half-assed! If they are, you risk having the .223 bullet strike the side of the spool tube, and since the tube will then be distorted and driven into the screen wire doughnut baffles in the front of the can, you'll be screwing up a lot of your hard work!

By contrast, the hole size (and spacing of said holes) in the sides of the 4 inch tube (or in the sides of the 1 1/8 inch O.D. 17 inch tube) are not critical. These holes are just there to vent gases, and the hot, pressurized powder gases will "find" these holes. So you needn't run out and buy the exact size drill bit (if you have the next size), and do not think you need to scrap the entire silencer just because you've misplaced a few vent holes by 1/16 inch.

In *How to Make a Silencer for a .22*, I said you could make the baffles for the front half of the can (3 inches worth in this .223 can) out of aluminum screen. That is still true. However, I recommend stainless steel screen wire for the rear 3 inches of baffles rather than screen wire made from bronze, copper, or brass. This is because the much greater heat of the .223's gases requires greater heat resistance than those materials offer.

The inside tube of the can (1 1/8 inch O.D.) is 1 inch longer than the outside tube (1 1/2 inch O.D.). This allows you to squeeze the slit ends of the can with the hose clamp. If the 1 1/2 inch (sleeve) tube were to reach back to the end of the 1 1/8 inch O.D. tube, you wouldn't be able to achieve the clamping effect of the screw-tightened hose clamp upon the neoprene disk and the barrel because clamping action on the outside tube obviously would not compress the inside tube.

The 1 inch thick threaded plug (welded inside the can) must be installed there by a technique known as "plug welding." You simply drill three 1/4 inch I.D. holes through the sides of the 1 1/8 inch tube, centered 6 1/2 inches away from the rear end of the tube in the same lateral plane around the can and spaced an equal distance from each other. Assuming that you've machined "true" the ends of all the tubes, you can use a square and a scribe to make a mark 6 1/2 inches from the rear end of the 1 1/8 inch O.D. tube, mount this tube in your lathe, and let the cutter scribe a mark around the circumference of the tube. Compute the circumference of a circle by multiplying "pi" (3.14) times the circle's diameter. So 1 1/8 inch O.D. x 3.14 = 3.53 inches circumference of the 1 1/8 inch O.D. tube. Use a flexible tape measure to mark off 1 1/8 inch from your first centerpunch mark, and make another punch mark. Repeat this, and then drill your three 1/4 inch I.D. holes. Deburr the insides of these holes (using a hook-shaped tool you've ground out of a three-sided deburring knife or a three-cornered file; see Fig. 10). Insert the 1 inch thick disk (which you've threaded to match the 7/8 inch x 14 thread of the adapter on your barrel's muzzle). Set your square to 6 inches and use it to align the centerline of the disk with the centerline of the 1/4 inch holes you've drilled in the 1 1/8 inch tube. Use your centerpunch to mark where you wish to also drill three holes (not too deep!) into the sides of the disk. Now you can use the MIG welder to permanently install this threaded disk into the 1 1/4 inch tube. Now you've got a tube that'll thread onto your Mini-14's barrel.

Wrap stainless steel screen wire around the spool of the expansion chamber and install this assembly into the can (from the front end of the 1 1/8 inch O.D. tube, of course). Do you see how the vent holes of the spool tube will permit the hot powder gases to exit the rear 4 inches of the 10 inch part of the can that's in front of the muzzle and flow into the area in between the two tubes and around the barrel?

Obtaining washers for welding in between the 1 1/8 and 1 1/2 inch tubes can be a pain, so make your own (Fig. 11). Drill a 1/16 inch hole in some .035 (35/1000) inch flat stock, use a kid's compass to draw the concentric circles, and use a coping saw with metal-cutting blade to cut them out. You'll need to drill 1/4 inch holes to start your saw cuts, of course, and you'll probably need to true up your washers with a Dremel hand-held grinder and carborundum cut-off disks. These washers needn't be perfect since the TIG welder can fill in small gaps. You must be gentle with the foot petal and that welding heat, however.

Now weld the washers into place. The area between the 1 1/8 inch O.D. tube and the 1 1/2 inch O.D. tube is now a sealed unit. Do not weld in these washers before you've "vented" the 1 1/8 inch tube.

While you can cut the snap ring groove at the front of the can with a Dremel and cut-off disks, it's best to perform this operation on the lathe. Make a hook-shaped cutter and make the groove .020-.030 (20/1000-30/1000) inch deep and .010 (10/1000) inch wider than the thickness of your 1 1/8 inch O.D. internal snap ring. Make certain that the bottom of this groove has a square corner on the front side in order to maximize its retaining effect against the gas pressures exerted on the end cap washer and the snap ring that holds the washer atop the baffles.

I now will describe the part that confuses everybody! If you carefully read and reread this description of how to crush-form the screen wire into the doughnut-shaped baffles of this design, you will understand how simple a process it actually is. Follow the process as depicted in the sketches and use a piece of paper spindled around a pencil to simulate the screen wire around one of the forming rods. Use a toilet tissue tube to simulate the cylinder of the baffle-forming die set. You'll then be able to see that the process I describe herein will result in a screen wire baffle that looks like the in-line "rock" filter found in your gas line at the carburetor.

Note that there are two different sizes of baffle (Fig. 12). The large one is approximately 5/8 inch long and has an I.D. of 5/8 inch. Since the 1 1/8 inch O.D. tube is .065 (65/1000) inch wall thickness, its I.D. is .995 (995/1000) inch, so .995 (995/1000) is the O.D. of all the baffles (both small and large). The small baffle is 3/8 inch long and has an I.D. of .250 (250/1000 or 1/4) inch. When assembled, there will be a 3/16 inch thick neoprene washer on the downrange side of each small baffle (Fig. 13). These neoprene wipes have an I.D. of .250 (250/1000, or 1/4) inch. These wipes help the can's efficiency by retarding the gases' linear flow somewhat, thus giving the screen wire mesh more time to dissipate the heat energy. By making the baffles in two sizes, I reduce the weight of the can, and by thus exposing more of the mesh to the gases, I increase its heat-sinking capability.

Note that the rear baffles are made of stainless steel screen wire while the remaining baffles are made of aluminum screen wire. The aluminum screen is lighter and is a better heat sink, but it cannot take the high levels of heat and blast present at the very rear of the can. It will melt! The stainless steel screen can take this heat.

Naturally, after you've drilled the 1/8 inch vent holes in the sides of the 1 1/8 inch tube, you must deburr them on the inside of the can so that the screen wire baffles can be slipped into place without snagging on these burrs. The larger the hole in a can's baffles, the more powder gas can follow the bullet out of the barrel, making noise. Thus, we want the minimum size hole commensurate with ensuring that the bullet does not nick a baffle. The .013 (13/1000) inch clearance used in this design achieves this desired effect fairly well. That means that there's .013 (13/1000) inch of air all around the bullet as it passes through the silencer. The tighter the tolerance between the I.D. of the baffle and the O.D. of the bullet, the straighter the can must be mounted on the barrel, naturally!

The .065 (65/1000) inch wall thickness of the 1 1/8 inch tube is just barely thick enough to allow the cutting of the .025 (25/1000) inch deep retaining groove in the front end of the silencer tube. The 1 1/8 inch O.D. snap ring is held in this groove, and the snap ring in turn holds the washer that holds the baffles in the can (Fig. 14). This arrangement will hold the normal working pressures of the can, but if something goes wrong, the snap ring will spring free of the groove and the baffles will be blown out of the silencer, but at least there will be no explosion as can happen with a solid, threaded front end plug. My way is also simpler, weighs less, allows the use of thinner (lighter) tubing, and wastes less precious space inside the can.

The powder gases are under pressure, so they will find the tiniest hole in the tube or a weld. The resulting leak will be as loud as if there were no silencer on the gun at all! It is very easy to burn a hole in the tube while welding in the washers in between the tubes at each end. This job requires that the welder know how to work with the minimum of amperage on his TIG machine and that he angle his arc so that 90 percent of the heat is directed away from the tubing! If the firing noise of your silenced .223 sounds louder than a whipcrack, immediately suspect a leak in the welds someplace (either a pinhole or a crack) or check at the slits and neoprene disk. Smear very soapy water on all joints, fire again, and watch for bubbles. If none are seen, repeat this process. If you still see no bubbles, carefully run the TIG welder over all joints again and test-fire again. The likelihood of trouble is a good reason to make two of all parts.

Now comes the part that many people find confusing: the making of the baffles out of screen wire. If you read through this description several times, referring to the respective illustrations on your second and third reading (and perhaps folding and spindling a paper rectangle around your pencil), I believe that you will see what I mean.

The five parts I will now refer to are not part of the silencer; they are part of a "forming die set"—a tool kit, if you will. This die is arm-and-mallet powered, so you needn't have a punch press or anything like that. I call these parts the piston, cylinder, washer, small punch, and large-punch. (I describe how to make these parts at the end of the text.)

Note that the punches, piston, and cylinder are all 5 inches long. The washer and the piston are slip-fits inside the cylinder. The little nipple of the large punch is a slip-fit in the washer (as is the small punch). The piston has two different sized holes in its ends—one is a slip-fit over the large punch; the other is a slip-fit over the small punch. The first is used to make large baffles; the second is for making small baffles.

Let's work through the making of a large baffle out of aluminum screen. Actually, you will need the stainless steel screen first when assembling the can, but stainless steel is hard to get (a source for all hardware is listed in the rear of this book), and I know that some of you will want to try out the baffle making before investing in the can parts or a threaded barrel!

Use a kid's pointed paper scissors to cut out a rectangle of screen, cutting with the weave to minimize the number of loose wire ends. The width of this screen rectangle will determine the length of the finished baffle (because the length of the screen rectangle is spindled around the punch). Therefore, the large baffle rectangle will be much shorter and wider than the rectangle for the small baffle. I can only give you approximate sizes, because different wire and mesh sizes will compress differently. You must experiment with what you have. Record the measurements of a test rectangle, and then make the next one larger or smaller as needed. The stainless steel screen compresses considerably tighter than the aluminum screen will, so the stainless rectangles will need to be larger than the aluminum rectangles.

Fold the screen rectangle lengthwise into thirds, as illustrated in Figure 15. One side will then have no cut ends. Fold one end of this (now folded) rectangle back on itself, with the loose ends toward each other (Fig. 16). Start spindling the screen at this fold, and then there'll be no loose wires in the tunnel through which the bullet must pass (Fig. 17). When finished, the screen wire baffle will look like a doughnut (like an in-line rock filter out of a car's carburetor, only with a hole in it for the bullet to pass through).

When you stack these doughnuts in the can, the bullet goes through the holes while the hot powder gases expand out into the screen wire baffles and become trapped. All the little wires soak up the gases' heat energy, leaving no blast of hot air to make noise. Once crushed to shape, the screen wire tends to not unravel, but make certain by dropping a bit of silver-solder on the appropriate spots (after forming the baffle and prior to slipping it into the actual silencer).

With the screen folded and spindled around the large punch, start the assembly (punch and screen) into one end of the cylinder. Once it's all inside, slip the die washer over the nipple of the punch. Put the nipple end of this assembly down on the bench (making certain that the washer is up inside the cylinder). Now slip the large-hole end of the piston down in the top of the cylinder and around the large punch. The screen wire is now trapped between the washer and the piston, with the rear of the piston protruding out of the top of the cylinder. With a rubber mallet, hit the top of the piston a good lick (about like you'd use to drive a large nail). This blow will "form" the screen wire into a doughnut-shaped baffle.

The length of the screen rectangle should be enough to require you to use the points of your scissors to tuck the ends of the screen into the cylinder when you're inserting the punch and screen for forming. So start with a long screen rectangle and trim it down as needed. After forming the large-hole baffle, measure its length. If it is less than 5/8 inch, make the screen rectangle wider. If the finished test baffle is longer than 5/8 inch, make the screen rectangle narrower. Once you get it right, record the rectangle's size for future reference.

Soft-solder won't adhere to aluminum or stainless steel, but the molten solder will become entangled in the screen wire's mesh and prevent the baffles from coming unraveled. After forming the baffle, remove the punch and washer and use a wooden dowel to push the baffle part way out of the cylinder, where you can put a small drop of solder on both the inner and outer ends of the rectangle. Push the baffle back toward the other end of the cylinder and repeat this soldering process at the other end of the baffle. Now remove the finished large-hole baffle from the cylinder and set it aside.

Now let's do a small-hole baffle. The screen rectangle will need to be considerably longer and narrower than the one for the large baffle. Note that the small punch has three "bumps" of weld on one end. These serve as a sort of bolt-head to keep the die washer from slipping off of the punch. Put the washer on the small punch before spindling the folded screen around the punch (Fig. 18). Insert the assembly into the cylinder as before, and drop the small hole end of the piston over the top of the small punch (Fig 19). Use the mallet and remove and measure the baffle. Once you have the size right, solder and set aside as you did the large baffle and record the rectangle size.

Annealing the stainless steel screen may be necessary if the screen is too "elastic" as it comes from the store. After folding the screen lengthwise, hold one end with a pliers and pass the screen through the flame of a gas kitchen stove. Once all of the screen has been made to glow red and allowed to cool, it will hold a formed shape much better. You can also anneal the screen with a propane torch or the acetylene (only!) flame of an oxyacetylene torch. Aluminum screen does not need to be annealed. When making the stainless steel screen baffles, remember to start with a larger rectangle, because the stainless screen compresses a lot more than the aluminum screen. When annealing, do not let the flame stay on any one part of the screen or you will burn a hole in it!

Now for the neoprene wipes. Industrial hardware supply houses stock neoprene in 4 x 8 foot sheets, like plywood. You can get a 3/16 inch thick, foot-square piece of it cheaply. It's used under the base of a bench-mounted grinder to reduce vibration.

Using the die set's piston as a template, draw circles on this neoprene. You'll need five of these wipes. Haggle them out with your pocketknife (taking care not to cut inside the circular pen marks). Pinch one between your fingers and apply the edge (held vertically) up to a bench grinder. Using the pen-drawn circle as a guide, go around the circle, grinding the neoprene to shape. After you're close to being done, taper one edge so that the wipe will start into the can more easily. Do not attempt to put the 1/4 inch holes in the wipes prior to assembling the can! There's no way to know which way the neoprene will compress when it's inside the can, so any premade hole will be off-center. We'll get that hole where it needs to be later.

Let's pack the baffles into the can. First, get a foot-long piece of the same 1/4 inch O.D. drill rod that you used to make the small punch of the baffle-forming die set. Turn in a lathe or grind one end of this rod into a long pencil point, flatten two sides of this point, and bend it into a buttonhook. Turn a rounded end on this 1/4 inch O.D. rod (Fig. 20). This end of the rod will be used to "burn" 1/4 inch I.D. holes in the neoprene wipes after a bullet has torn holes where they need to be. The buttonhook is for reaching into the can and removing baffles without damaging them. It's best to allow for mistakes up-front, right?

A large stainless steel screen baffle goes into the can in front of the expansion chamber. Use the point of the scissors to work the slightly oversize baffle into the can, going around its circumference, tucking the screen edges in much as you would work a bicycle tire onto its rim. Use a dowel to push this first baffle all the way into the rear of the can. Follow it with a small stainless steel baffle and then with a 3/16 inch neoprene disk. Repeat this (two each of large and small stainless baffles), then insert a large-hole aluminum screen baffle, then a small-hole aluminum screen baffle, then another neoprene disk. Repeat this until your can is packed (Fig 21). Insert the end cap washer (Fig. 22; be sure you don't install the die washer accidentally!) and install the 1 1/8 inch internal snap ring to hold everything in place. (Your paper scissor's points can be ground to use as snap ring pliers, saving yourself $15.)

The length of the baffles can be varied a little, if need be. If there's 1/8 inch too much packing in the can to let you install the end cap washer, you can get away with using the piston and a mallet to smash things a little tighter and to get the snap ring in place (temporarily leave out the end wipe to remove some of the "bounce"). Do not count on much of this sort of thing to cover up for shoddy workmanship, however. This .223 silencer in particular requires meticulous attention to detail or you'll end up with nothing but scrap. In view of all the trouble you've gone to, best not endanger the entire project by cutting corners somewhere.

You're now ready to test-fire your new .223 silencer! It takes a lot of wood to safely stop a .223 round, so get an oak 6 x 6, or use eight pine 2 x 6s or a 5-gallon bucket of sand underneath a 2 x 6 board. Load *one* round into your .223's chamber. *Do not load the magazine!* Ideally, use the .22 LR conversion unit for this first test, then the .223 later.

Carefully touch the can's muzzle (held vertically) to the 2 x 6 and fire down into the board. Holding everything steady, raise the gun an inch and move it down enough so you can observe the bullet hole in the wood. Are your washer and snap ring laying there on the board? If so, you have a problem! Ditto if the bullet hole is elongated. Look at the front of the can. Is the washer knocked loose? Is there a copper or lead smear on one side of the hole in the end cap washer? If so, perhaps you can tell which way to "bend" the can with a light rap from your rubber mallet.

If the bullet hole is round and all is well, heat the end of the 1/4 inch foot-long rod red-hot with your torch and pass it through the neoprene wipes, pausing at each one long enough to feel the heated rod poke through the wipe. This operation will not quite be a continuous push—there will be momentary pauses as the hot end of the rod cauterizes the neoprene, converting the raggedly torn bullet hole into a nice, round, centered hole.

If your bullet hole was elongated but the washer remained in place, don't panic. Just burn the holes in the wipes and try another round. Occasionally, the five hunks of neoprene are enough to upset the slug (which the barrel's rifling has caused to spin on its axis). Once you have those 1/4 inch holes in the wipes, though, the slugs should fly true and the hole in the board should be round. If it is not, try using the 1/4 inch rod to "wallow" any misplaced screen wire out of the bullet's path, and fire another round. If you've still got problems, check to make sure the gun is unloaded; then, leaving the can on the barrel, look through it while shining a goose-necked bore light from the chamber. Look for misplaced baffle material. If you still can't see where the problem lies, carefully unpack the baffles, looking at each one to see if it is unraveled. If so, make more, repack the can, and try again.

These baffles do not "burn out" as does steel wool, fiberglass insulation, and other crap you'll see touted for use in silencers. The screen wire will, however, fill up with carbon residues and powder grains. As the baffles fill up, they can't do their job anymore, and the gun will get progressively louder with each shot. Your can should be good for at least 300 rounds (possibly as high as 1,000) before needing cleaning or baffle replacement. If you must repack it after every magazine you fire, how can you have a tested silencer and any reserve usefulness left for actual recreational use?

You can either replace the carboned-up baffles or clean

them out and reuse them. They'll come cleaner if you remove them, naturally. Let them soak in carburetor cleaning solvent and blow them out with an air nozzle.

It is possible to do a credible cleaning job while leaving the baffles in the can. Remember, unless something makes removal necessary, leave the can on the barrel! You can use gasoline as a solvent, too, but be sure it's all blown out and evaporated before firing another round through the can. A silencer fire is no joke! It's pretty easy to have spare baffles and neoprene disks made up in advance.

When a semiautomatic firearm cycles, a lot of noise comes out of the ejection port with the empty cartridge casing. If you'd like to see just how much extra noise this is, C-clamp a leather pad in such a way as to lock the bolt shut. The bullet's impact can also make quite a bit of noise, especially if it's hitting metal. If you'd like to see how much noise bullet impact is adding to your testing, put some folded rags on top of your boards, or fire into sand, dirt, or (if you're in a safe area) up into the sky. The .223's velocity exceeds the speed of sound, so it causes sonic crack. Firing straight up into the sky can sometimes prevent the crack from being heard! Try firing at night, holding the piece so that you can observe the front of the can. There should be no muzzle flash at all. If there is, your can is too short, not well-packed, or needs cleaning.

I've left until last the description of how to make the baffle-forming die set. That was done for a reason. People seem to get the idea that the piston, cylinder, and punches are somehow part of the silencer itself. They are not. They comprise a simple tool kit that can be used over and over to make baffles for many silencers.

As I said before, the cylinder must be the same I.D. as your can. Simply cut off a 5 inch long piece of the same 1 1/8 inch .065 (65/1000) inch wall tubing you'll use to make your can, true its ends in the lathe, and deburr it (Fig. 23).

The piston is more complicated (Fig. 24). You can either turn it out of solid round stock on the lathe or make it out of steel tubing and weld washers in each end. The piston's O.D. needs to be .005 to .010 (5/1000-10/1000) inch smaller than the I.D. of your cylinder. If you're using the 1 1/8 inch .065 (65/1000) cylinder, you'd want your piston's O.D. to measure .985-.990 (985/1000-990/1000) inch. This will allow it to be a slip-fit inside the cylinder. The holes in the ends of the piston need to measure .255-.260 (255/1000-260/1000) and .630-.635 (630/1000-635/1000) inch, respectively, in order to be slip-fits over the small or large punches (whichever you're using to make small or large baffles).

The die washer needs to have an O.D. of .985-.990 (985/1000-990/1000) inch (just like the piston). The washer's I.D. should measure .255-.260 (255/1000-260/1000) inch so that it is a slip-fit on the small punch and on the nipple of the large punch (Fig. 25). If you must grind and case-harden an Allen wrench for use as a small boring bar to enlarge the I.D. of a washer, do so (Fig. 26). Bore out the I.D. of the washer after welding it onto the end of the piston.

I use an Allen wrench to form the cutting tool for making the internal snap ring groove at the front of the can, too. I simply hold the Allen wrench in a bored-out piece of round stock by means of a set screw and lock this bar in the tool post of the lathe.

The large and small punches are shown in Figures 27 and 28. Naturally, you will use the lathe to produce the 1/4 x 3/8 inch long nipple on the large punch and perhaps the shaped ends of the foot-long 1/4 inch O.D. rod that is used to burn the holes in the neoprene wipes. You will probably need the lathe to reduce the O.D. of the piston in order to make a slip-fit inside the cylinder. Of course, you will need to deburr all the sharp corners of the baffle-forming die set as well as the two can tubes, the end cap washer, the two sleeve washers, and the rear end plug.

I hope I've explained everything well enough. As I said before, this book is not for those who are not machine-shop trained, nor for those who are unfamiliar with the Mini-14's inner workings, nor for those who haven't read *How to Make a Silencer for a .22.* If you ahve any questions about the material in this book, or about silencers in general, I refer you to that earlier book. Take care, and be quiet!

LIST OF SUPPLIERS

NOTE: All addresses were current at the time of this writing.

BUREAU OF ALCOHOL, TOBACCO, AND FIREARMS
(BATF, or ATF). Look in the white pages of your phone book under "Government, Federal."

B-Square Company
Box 11281
Fort Worth, TX 76110
(scope mounts)

Brownell's Supply, Inc.
200 S. Front St.
Montezuma, IA 50171
(gunsmith tools and advice)

Federal Cartridge Company
900 Ehlen Drive
Anoka, MN 55303-7503
(.62 ball ammo, .69 match ammo)

L.L. Baston Co.
Box 1995
El Dorado, AR 71731
(.22 conversion units and folding stocks)

McMaster-Carr
Box 4355
Chicago, IL 60680-4355
(all hardware needs—screen wire, tubing, tools, neoprene, etc.)

Midway Reloading Supplies
5875 W. Van Horn Tavern Road
Columbia, MO 65203
(.22 ammo)

Olin Winchester
E. Alton, IL 62024
(64-grain softpoints)

Sturm Ruger & Company
Lacey Place
Southport, CT 06490
(Mini-14 ports, magazines, folding buttstocks)

Trijicon Inc.
Box 2130
Farmington Hills, MI 48333
(luminous sight inserts)

USA Magazines Inc.
P.O. Box 39115
Downey, CA 90239
(magazines)

Williams Trigger Specialities
Rt. 1, Box 26 A
White Heath, IL 61884
(trigger jobs)

FIGURE 1

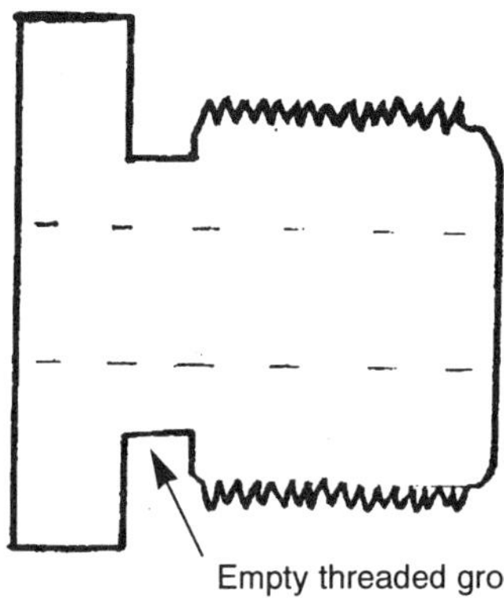

7/8" x 14 TPI bolt (1" of threads)

Precision bored and honed hole .001" smaller than O.D. of barrel.

Allow for barrel taper!

This adapter is a press-fit onto the barrel (nonremovable).

FIGURE 2

Plug weld (with MIG wire feeder) through the three 1/4" I.D. holes in 17" tube and into matching holes in threaded plug.

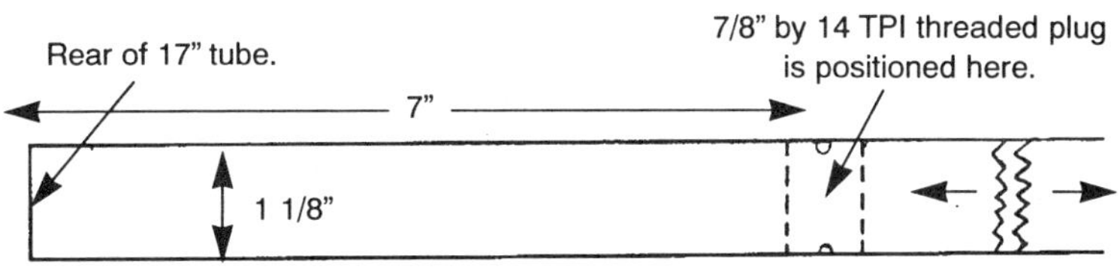

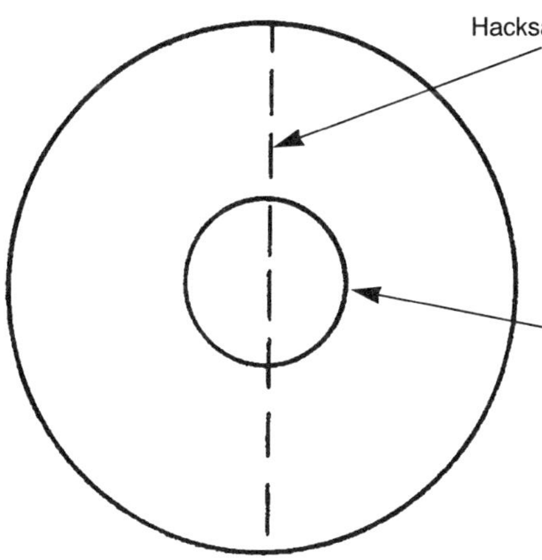

FIGURE 3

5/8" thick neoprene.

I.D. is 3/32" smaller than the O.D. of barrel at forend of stock.

O.D. of neoprene is 3/32" larger than I.D. of 17" tube.

FIGURE 4

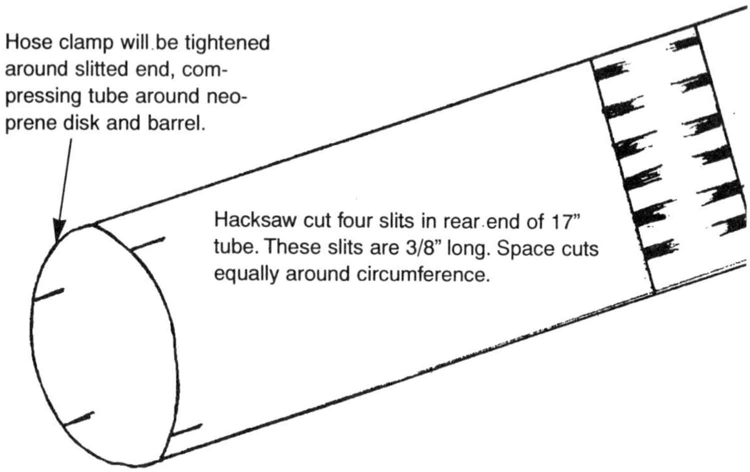

FIGURE 5

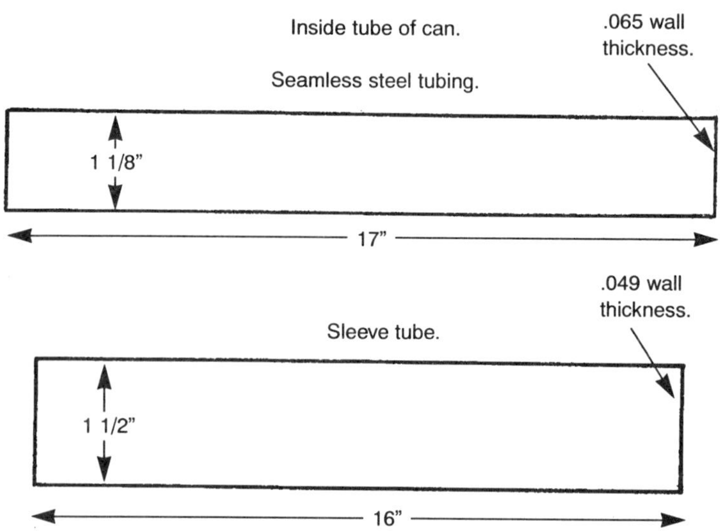

FIGURE 6
Expansion chamber.

Tolerances +/– .001" on .990" and on .502".

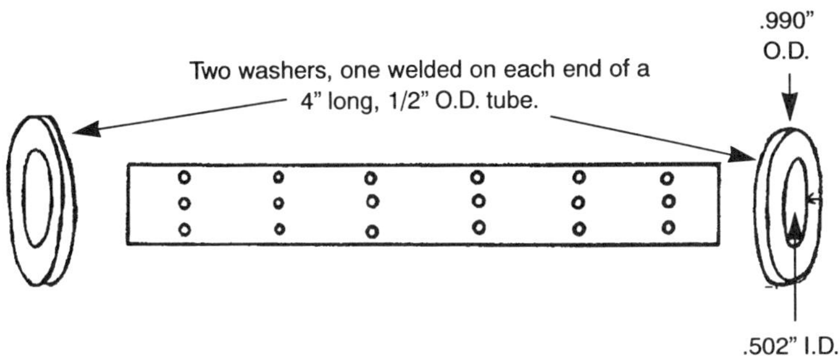

FIGURE 7
Expansion chamber.

Rows of 1/16" holes drilled into tubes. Four holes around circumference.
Stainless steel screen wire is *loosely* spindled around this 4" tube.

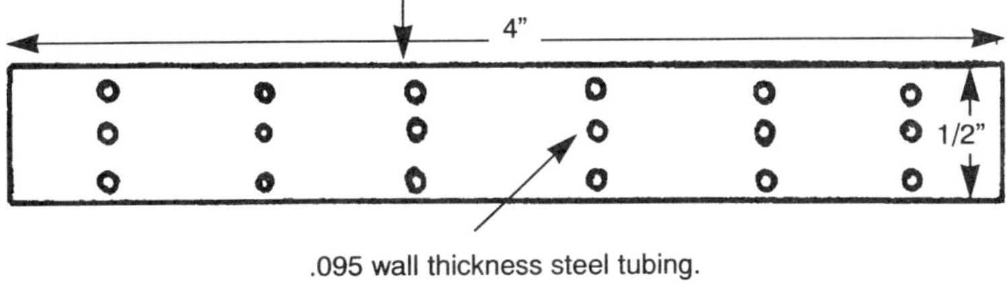

FIGURE 8
17" tube (1 1/8" O.D.).

Rows of 3/16" I.D. holes, eight holes around the circumference.

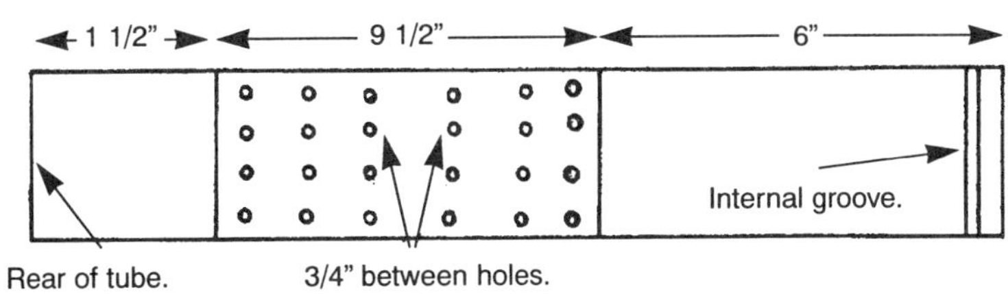

FIGURE 9
Welded-in plug.

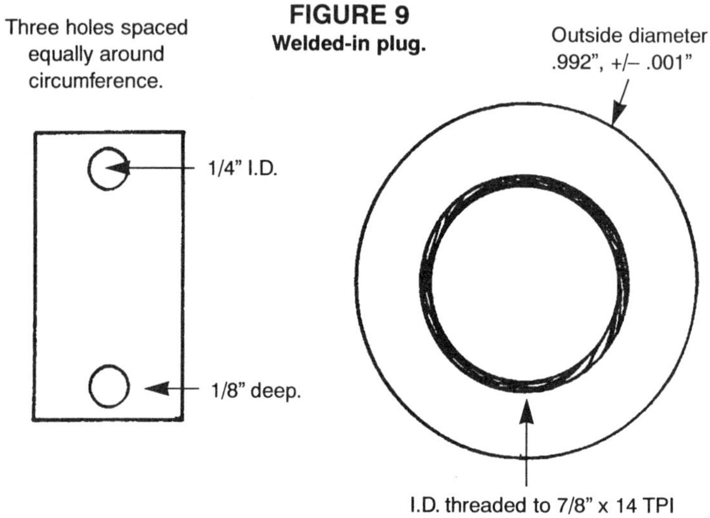

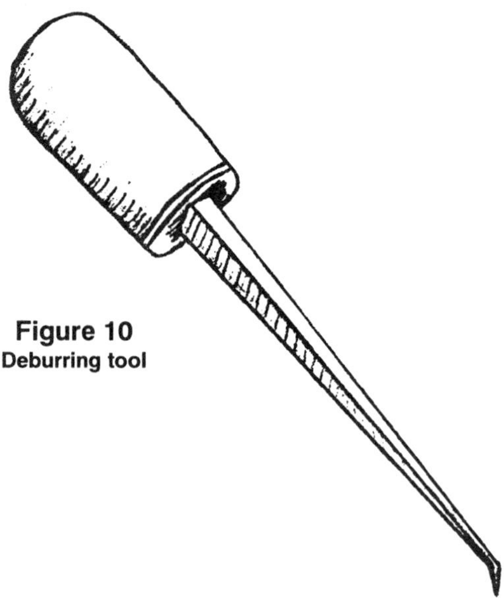

Figure 10
Deburring tool

FIGURE 11

Making sleeve washer.

Drill access holes on the edge and in the center.

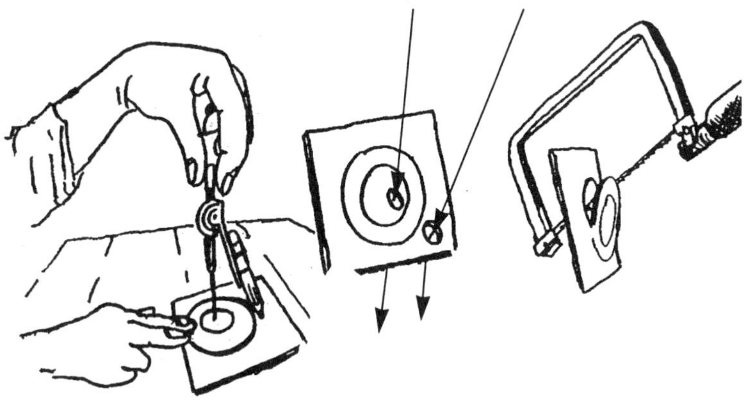

FIGURE 12

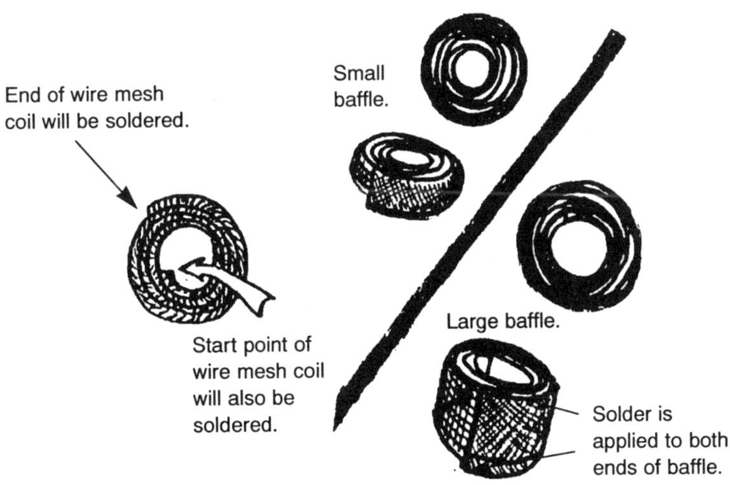

FIGURE 13
Neoprene wipe.

Important: Do *not* attempt to fashion center holes in wipes prior to assembly in the silencer!

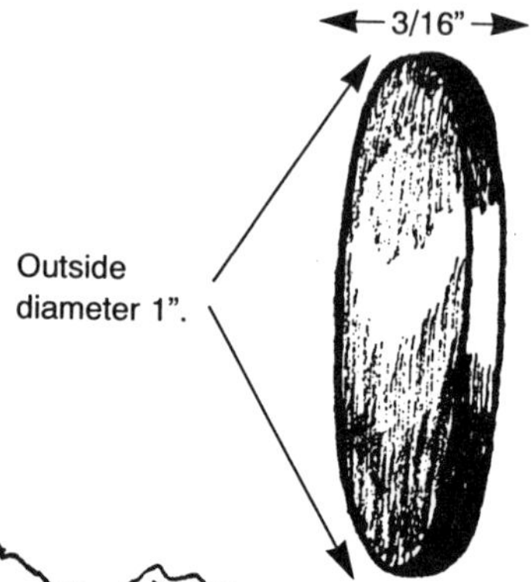

FIGURE 14
Inserting snap ring.

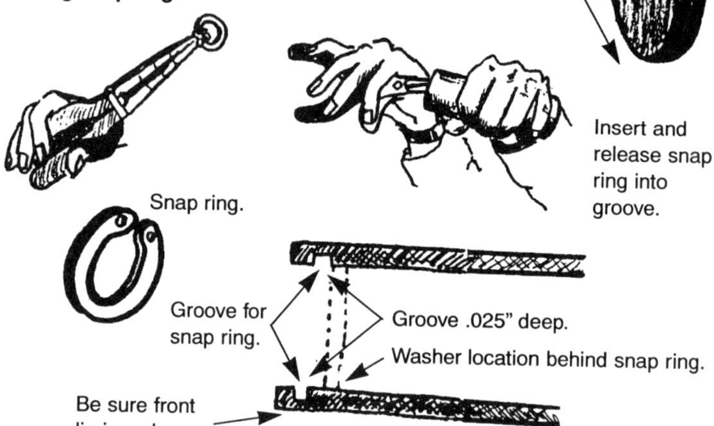

FIGURE 15
Screen wire mesh sheet.

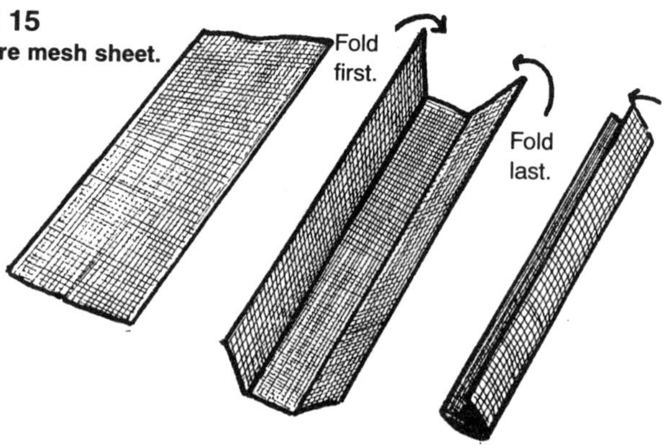

FIGURE 16

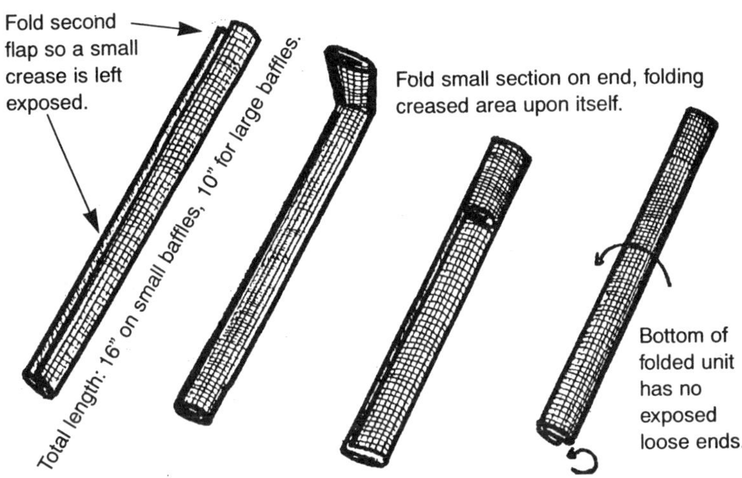

FIGURE 17
Spindling using large punch.

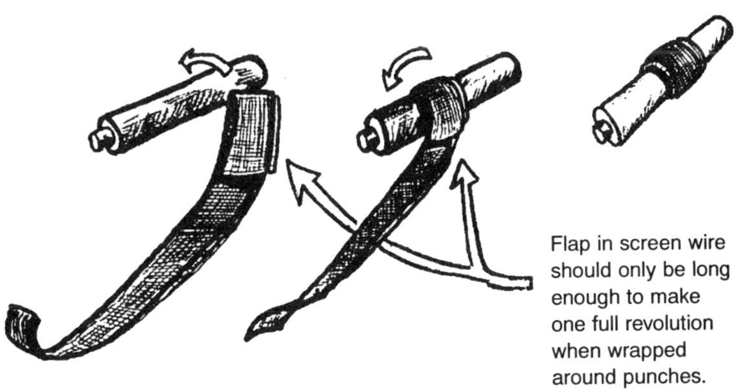

FIGURE 18
Spindling using small punch.

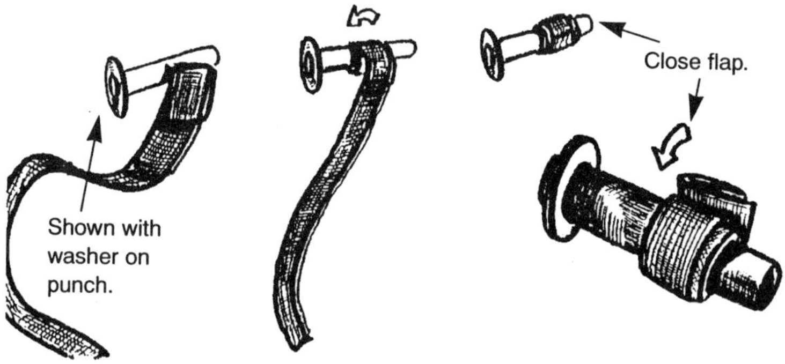

FIGURE 19

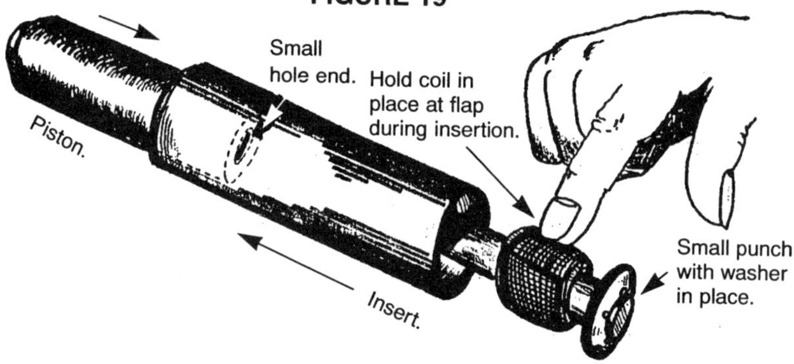

FIGURE 20

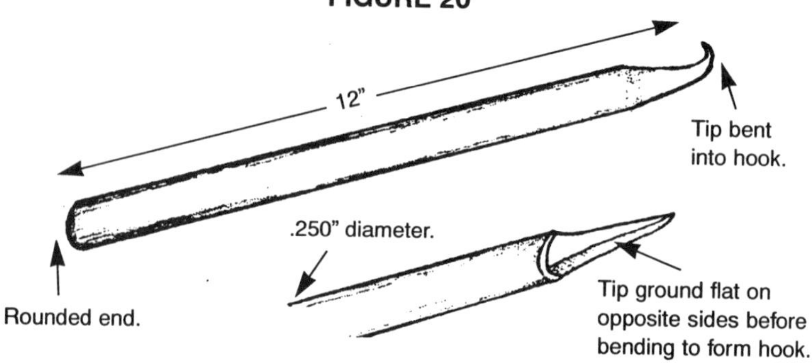

FIGURE 21

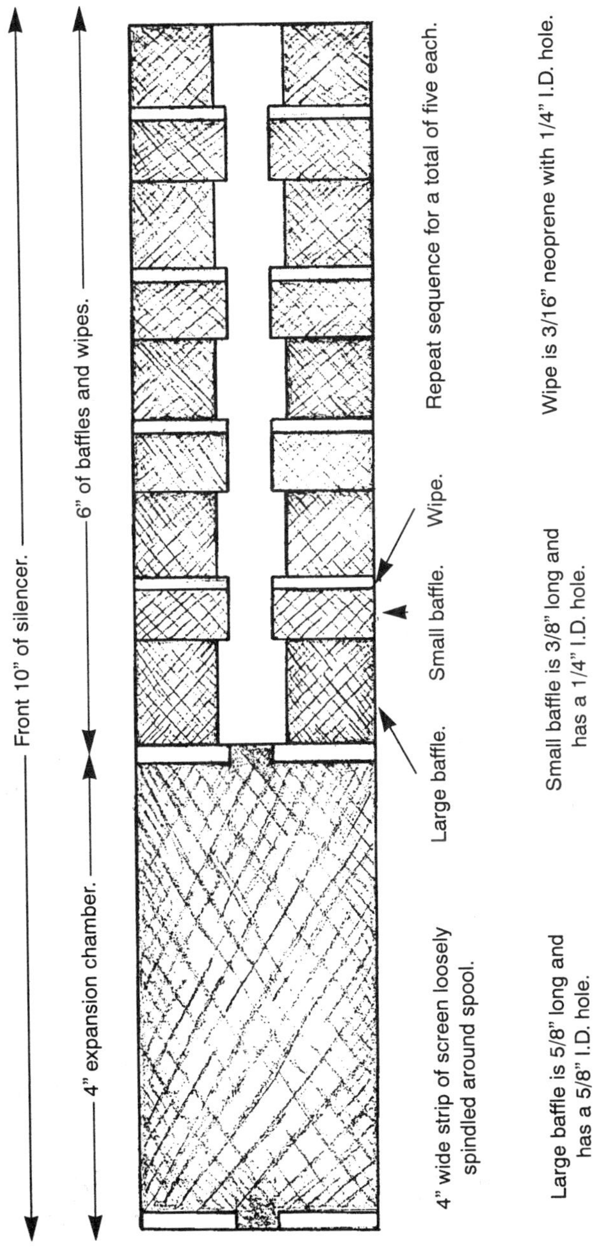
Front 10" of silencer.
6" of baffles and wipes.
4" expansion chamber.
Repeat sequence for a total of five each.
Wipe.
Small baffle.
Large baffle.
4" wide strip of screen loosely spindled around spool.
Wipe is 3/16" neoprene with 1/4" I.D. hole.
Small baffle is 3/8" long and has a 1/4" I.D. hole.
Large baffle is 5/8" long and has a 5/8" I.D. hole.

FIGURE 22
End cap washer.

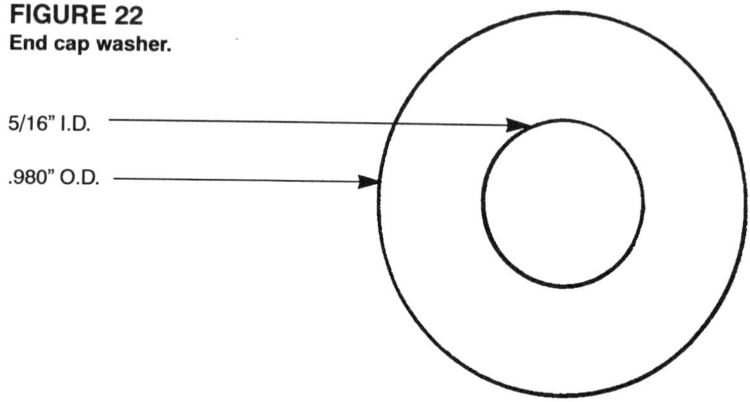

FIGURE 23
Cylinder.

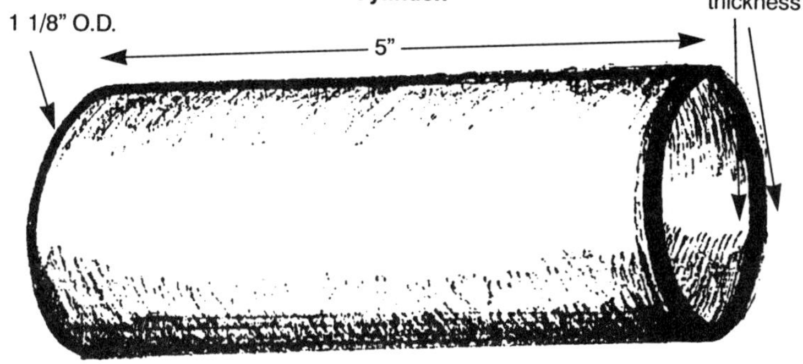

FIGURE 24
Piston.

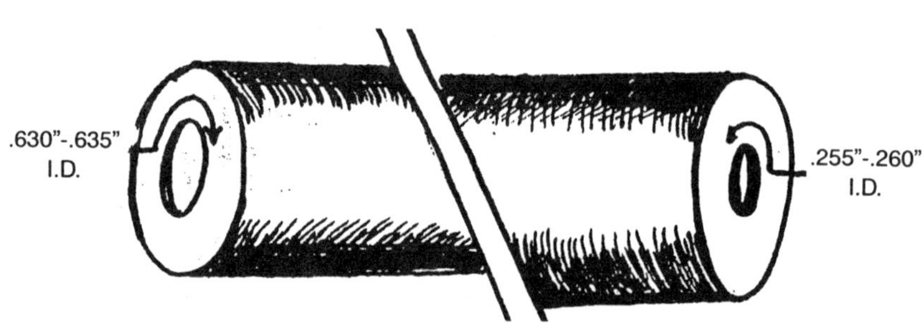

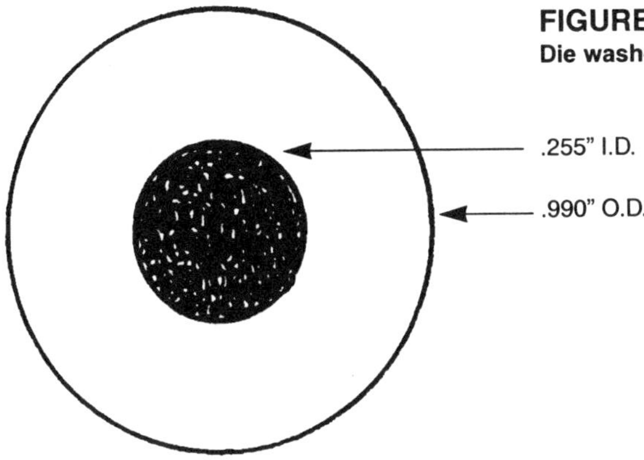

FIGURE 25
Die washer.

FIGURE 26

Allen wrench ground to shape and case-hardened for cutting snap ring groove.

FIGURE 27
Large punch.

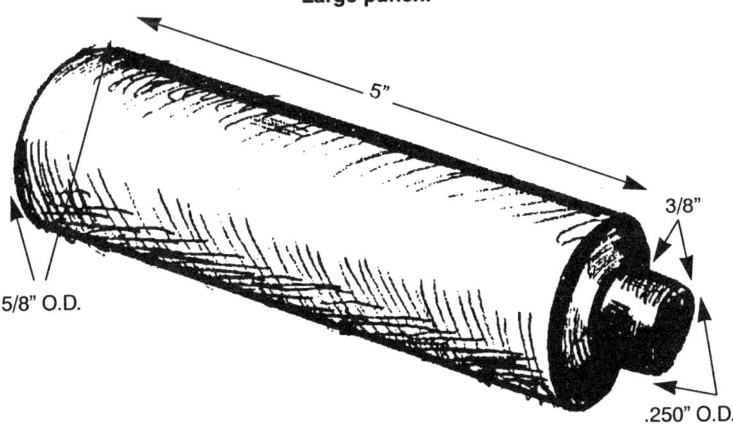

FIGURE 28
Small punch.

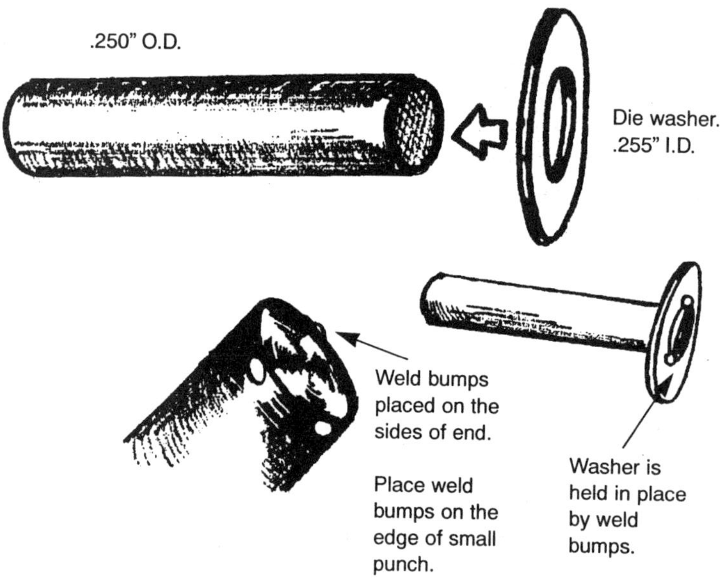